SPILL!

SPILL!

The Story of the Exxon Valdez

Terry Carr

FRANKLIN WATTS 1991
New York • London • Toronto • Sydney

To my mother and father

Cover: This dead deer found floating off Knight Island, Alaska, may have died from eating oil-soaked kelp.
Frontispiece: Oil drips from the gloves of a cleanup worker on the beach of Green Island, one of the areas hit hardest by the oil spill.

Maps by Vantage Art, Inc.

Photographs courtesy of: Gamma-Liaison/Anchorage Daily News: pp. 14 bottom, 18, 29 top left, 36 bottom, 60 (all Erik Hill), 23 (Michael Penn), 33 top left, 42, 44, 46 right (all Bob Hallinen), 45 bottom right (Nichols), 57 right (Jim Lavrakus), 8, 9, 11 left, 26, 32 top, 52, 54, 57 left; Gamma–Liaison: pp. 2, 33 right, 34, 49 bottom, 51, 58, 59 (all Karen Jettmar), 20, 25, 29 bottom right, 31, 35, 38 bottom, 45 top right, 45 bottom left (all Ron Levy), 24 bottom left (Paquet/Rhapsody), 27 (Michelle Barnes), 39 (Seattle Times), 53 (Earl Cryer); Charles N. Ehler/National Oceanic and Atmospheric Administration: pp. 6, 29 bottom left, 30, 41, 50 left; David W. Johnson: pp. 11 right, 14 top, 22, 24 top and bottom right, 29 top right, 32 bottom, 33 bottom left, 36 top, 40, 45 top left and center, 47, 48, 49 top, 50 right, 61; U.S. Fish and Wildlife Service: pp. 38 top right; Center for Marine Conservation: pp. 38 top left (Meg Berlin), 46 left (Burr Heneman).

Library of Congress Cataloging in Publication Data:

Carr, Terry.
    Spill! : the story of the Exxon Valdez / by Terry Carr.
        p.   cm.
    Includes bibliographical references and index.
    Discusses the carelessness and neglect that led to the oil spill in Prince William Sound, Alaska; the cleanup effort; and the long-term consequences of this disaster.
        ISBN 0-531-15217-0. —ISBN 0-531-10998-4 (lib. bdg.)
    1. Oil spills—Environmental aspects—Alaska—Prince William Sound Region—Juvenile literature. 2. Tankers—Accidents—Environmental aspects—Alaska—Prince William Sound Region—Juvenile literature. 3. Exxon Valdez (Ship)—Juvenile literature.
[1. Oil spills—Alaska–Prince William Sound Region. 2. Tankers—Accidents. 3. Exxon Valdez (Ship)] I. Title.
TD427.P4C367 1991
363.73'82'097983—dc20   90-13104   CIP   AC

PmL

CONTENTS

7/31/91

APR 13

The Alyeska oil terminal in Port Valdez, in the spring of 1989

1

COLLISION

On the windless night of March 23, 1989, a light snowy mist settled over Port Valdez on Alaska's southern coast. From the town of Valdez, snuggled deep in the port, the lights of boats on the water hung like faint twinkles in the haze. A ring of mountains protects both the town and the port, and on this spring night the water lay flat and calm. The people of Valdez were turning in for the night.

Across the water from Valdez, at the Valdez pipeline terminal, the job of loading the oil tanker *Exxon Valdez* ended. About 53 million gallons (200,605 kl) of oil had been pumped into the huge vessel, which is owned by the Exxon Corporation, the nation's biggest oil company. It was the tanker's twenty-eighth visit to the terminal since 1986, when the vessel had been built.

At 9:26 P.M., the *Exxon Valdez* pulled away from the pipeline terminal. The plan was to steam south away from Alaska, travel along the western coasts of Canada and the United States, and deliver the oil to Long Beach, California. The trip would take about five and a half days.

The tanker was under the command of Captain Joseph Hazelwood, regarded as one of the best commanders of the Exxon fleet. The ship, which cost about $125 million to build, was one of the newest and finest of the fleet.

Tankers load millions of gallons of crude oil at the Valdez pipeline terminal.

Since 1977, when the trans-Alaska oil pipeline was completed, almost 9,000 tankers had loaded oil at the Valdez pipeline terminal and headed south, just as the *Exxon Valdez* was doing. The pipeline snakes 800 miles (1,287 km) across Alaska before ending in Valdez. This pipeline is the only means of getting oil from remote regions of Alaska.

The pipeline and the tankers had always posed a danger to the town, the water, and the nearby shorelines. One mistake could smear them with ugly, black oil. The area is one of the most beautiful in Alaska, but an oil spill could quickly change that. Despite this threat, oil development in Alaska had done good things for Valdez.

The **Exxon Valdez,** *shown here under tow, carried 53 million gallons (200 million l) of oil on the night of its ill-fated voyage.*

Before the 1970s, Valdez (pronounced Val-DEEZ) was a small, quiet town. Its economy depended on fishing and shipping. Then, in 1968, the Atlantic Richfield Company discovered oil hundreds of miles away, on Alaska's North Slope. It was a huge discovery that generated great excitement. Oil companies had been studying the North Slope for years as a possible source of oil. Atlantic Richfield's test wells confirmed that it was indeed a land rich in "black gold." More oil was there than in any other place in North America. Oil had made other states and countries wealthy. Alaskans thought it could do the same for them.

The state of Alaska, which owned the land over the oil, held an auction at which oil companies bid for the right to drill for oil. In 1969, the state sold $900 million worth of drilling rights to several oil companies in the first big lease sale of Alaskan land.

But problems arose. Alaska's North Slope is one of the coldest, most desolate regions in the world. How would the oil be drilled? How could the oil reach the rest of the United States? How would oil wells, roads, construction, and people affect Alaska's wildlife and environment?

State officials, the oil industry, and others debated these questions for years. One side embraced the economic benefits. They saw untold riches in the oil discovery. They imagined the state of Alaska making money from leasing land to the oil companies, from taxes on the oil companies, and from the share of the oil, called a royalty, that the state would receive. This side argued that the country needed Alaska's oil.

The other side worried about the risks. They argued that construction and the possibility of accidents were unacceptable dangers. They emphasized the hazards of oil spills to land and water. Many others were undecided. They wanted the wealth that oil would bring, but they wanted to protect the environment too.

Finally, in 1973, the United States Congress and President Richard M. Nixon approved construction of a pipeline from the North Slope to Valdez.

(Left) One-fourth of the nation's oil travels through the trans-Alaska pipeline, which ends at the port of Valdez. (Right) The Alyeska oil terminal supplies about 2 million barrels of oil a day to the rest of the United States.

The plan was to pump two million barrels (318,000 kl) of oil a day through a 48-inch (1.2 m) pipe to Valdez, where it would be loaded aboard tankers and shipped to the rest of the United States.

The line was called the trans-Alaska oil pipeline. It was 799 miles (1,294 km) long and was one of the most difficult and most expensive private construction jobs in history. It cost more than $9 billion, which was paid by the oil companies that had bought leases on the North Slope. Men and women from all over the country worked on the line. They endured great hardships, working far away from any towns in some of the coldest weather on earth.

The temperature sometimes sank to −60°F (−51°). Parts of the pipeline had to be buried. Parts had to be built on stilts above the ground. Other parts had to be built on bridges over rivers. A construction project like this had never before been attempted under such conditions.

When the job was finished in 1977, oil began to flow.

Some of that oil was now in the cargo tanks of the *Exxon Valdez*, which was making its way through Port Valdez. Aboard the tanker, the harbor pilot steered the vessel from the control room. Pilots differ from captains. Although the captain takes overall responsibility for the vessel, a pilot goes aboard to guide the ship until it leaves Port Valdez and is well on its way to open water. The pilot has studied the hazards of the region. He is familiar with any rocks and reefs. He knows that sudden changes in weather can ambush a tanker.

The *Exxon Valdez* is among the largest vessels on water anywhere. Almost a thousand feet (304 m) long, it plows the water with tremendous force. While moving at her top speed of about 15 miles (24 km) an hour, the ship takes 3 miles (4.8 km) to stop.

A little over an hour after leaving the pipeline terminal, the *Exxon Valdez* passed through Valdez Narrows. This narrow opening connects Port Valdez with Prince William Sound. Pilots and captains consider Valdez Narrows one of the most dangerous places on the tanker route. Ships must slow to about six miles (9.6 km) an hour. In the middle of Valdez Narrows sits a huge rock, called Middle Rock. Tankers must pass the rock through a channel less than a mile (1.6 km) wide. A sudden change in weather, engine trouble, or water turbulence can smash a ship against the rock or the shore.

On this trip, however, the pilot steered the ship safely past the rock.

Once into Prince William Sound, the *Exxon Valdez* gathered speed. The worst danger seemed to be behind her. Ahead lay the run to Hinchinbrook Island and Montague Island. These two islands guard the entrance to the Sound. The gap between them leads out to the Pacific Ocean.

At about 11:35 P.M., the tanker passed Rocky Point. Pilot Ed Murphy got off. He boarded a small boat and returned to Valdez. The tanker was now back in the control of Captain Hazelwood.

In the Sound, a drizzle was falling, but the fog had thinned. A mild wind blew from the east at about 10 miles (16 km) an hour. The trip ahead looked easy and trouble free.

At about this time, Captain Hazelwood radioed the Coast Guard station in Valdez. The Coast Guard is responsible for watching tanker traffic on radar and for controlling the ships' movements. When Captain Hazelwood made contact with the Coast Guard, he said he was changing course slightly to avoid icebergs that had drifted into his path from Columbia Glacier.

Columbia Glacier is one of more than 150 glaciers clinging to the mountains of Prince William Sound. The glaciers are left over from an ice age that existed about a million years ago. The earth's temperature dropped and ice blanketed almost the entire Prince William Sound region. Then, 10,000 to 15,000 years ago the temperature began rising again and the ice began melting. Some of the ice remains today in the form of glaciers, though many of these slowly continue to melt.

Columbia Glacier is far up in a bay on the north end of the Sound. The Columbia is continually losing ice, dropping great chunks into the water in a process called calving. Once in the water, these chunks become icebergs that float out into the Sound. The floating ice can be a danger to navigation.

An iceberg, unless it were a gigantic one, probably would not harm something as large as an oil tanker. But because most of an iceberg is hidden below the water, ship captains do not know for sure how much damage a drifting iceberg can cause. To be on the safe side, captains often steer their tankers away from the floating ice.

The tanker route through the Sound is like a divided highway. The lane on one side is for ships traveling south. The other lane is for northbound ships. A separation zone, a sort of median strip, divides the two lanes. When

Captain Hazelwood reported that he was changing course because of the floating ice, he meant he was going to swing left across the separation zone to the other lane. The Coast Guard radioed that no other tankers were on the "highway," so it was all right for the *Exxon Valdez* to cross the zone.

In the tanker's control room, Captain Hazelwood instructed his third mate, Gregory Cousins, to steer the tanker back into the correct shipping lane once the tanker reached Busby Island, a few minutes ahead. Then the captain went below to his cabin.

A few minutes later, Third Mate Cousins gave the order to the helmsman, who steers the vessel, to turn the tanker to the right. This turn would begin taking the ship back into the correct shipping lane. Another few minutes passed. Third Mate Cousins gave the order for another turn.

Something was seriously wrong. The vessel was not turning sharply enough. Either Third Mate Cousins had given the orders to turn too late, the helmsman had not carried out the command properly, or something was wrong with the way the ship was responding.

Critical seconds passed. It was now about midnight. The tanker had crossed the other lane and was heading toward a submerged rock called Bligh Reef. On deck the tanker's lookout, watching for channel markers and other lights, warned that the light marking Bligh Reef was to the right of the tanker, not to the left, where it would be if the ship was in safe water.

The tanker was turning, but it was too late. The third mate phoned Captain Hazelwood, who was still in his cabin, telling him, "I think we're in serious trouble." Immediately after he said the words, those aboard felt the first impact of the tanker with Bligh Reef. It was 12:04 A.M., Friday, March 24, 1989.

(Top) Ice breaks off from glaciers or icebergs in a process called calving. (Bottom) Icebergs calved from Columbia Glacier drift into the Sound, creating obstacles for ships passing through the area.

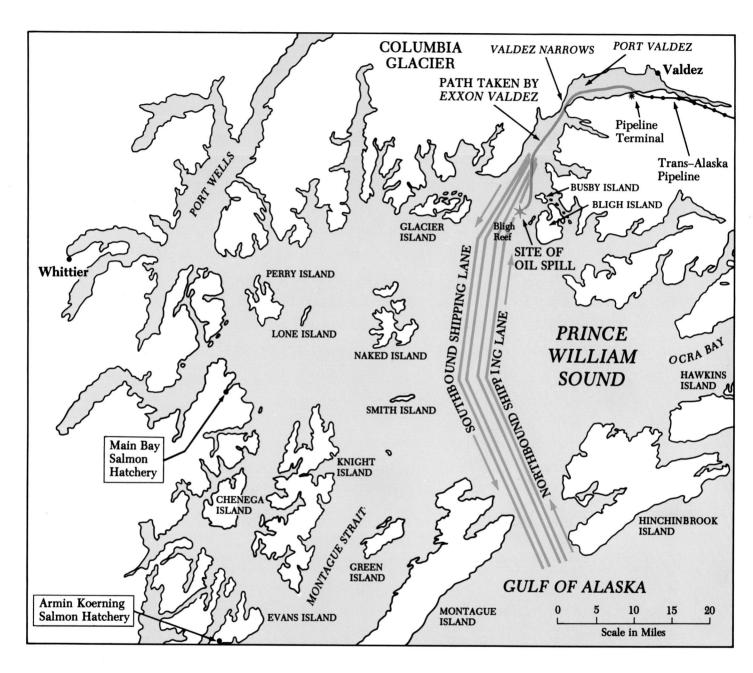

The route tankers take through Prince William Sound is like a divided highway. After the Exxon Valdez crossed over from the outbound to the inbound shipping lane, it failed to return to its normal course and struck Bligh Reef. The spill threatened salmon hatcheries only weeks before millions of salmon fry were expected to return to the Sound.

The tremendous impact of the tanker hitting the reef ripped open her steel cargo holds as if they were plastic. Tons of oil poured out, so fast and with such force that it created a wave of oil three feet high on the water. Once in the water, the oil fanned out, spreading like thick black ink in a bathtub.

The *Exxon Valdez* was stuck fast on the reef. The crew hurried to assess the damage and to find out how much oil had been lost. Fumes from the spilled oil rose up and covered the tanker. One crewman later said the odor almost knocked him out.

The sudden release of so much cargo threw the tanker out of balance. If the captain was not careful, the vessel could break in half or tip over. Nobody had been injured in the collision with Bligh Reef, but if the vessel capsized, the crew would not survive in the cold water or the oil.

Captain Hazelwood did not radio the Coast Guard about what had happened for more than twenty minutes. When he did, he told the Coast Guard, "We've fetched up hard aground. . . . Evidently we're leaking some oil and we're going to be here for a while." He said he was trying to work the tanker free from the reef.

At about 2:00 A.M., a crewman radioed the Coast Guard that the captain had given up maneuvering the tanker. The ship's engines were stopped.

Captain Hazelwood's actions that night would later spark heated arguments. First was the issue of his maneuvering the tanker after it hit the rock. Some criticized him for trying to free the tanker because such movements could do more damage to the ship. Others, however, praised him for his maneuvers because the ship was kept balanced and afloat. Second was the question of whether or not he had been drinking. At a National Transportation Safety Board hearing held after the accident, several witnesses testified that they could smell alcohol on the captain's breath before and after the accident. Others, however, said his judgment did not appear impaired.

But while these arguments would rage later, the immediate problem after the wreck was dealing with the spilling oil. Almost all of the tanker's

*The **Exxon Valdez** lies aground on Bligh Reef, with eight of its thirteen cargo tanks ripped open. Within a week its oil slick had blanketed a 900-square-mile (2,331-square-km) area.*

hull had been damaged. It was amazing that she remained afloat. The reef had peeled open the bottom of the *Exxon Valdez* the way a can opener would a can. One hole in the vessel was 18 feet (5.5 m) wide. Cracks ran nearly the length of the ship. For several hours, oil continued to pour from the holes and cracks.

In all, 11 million gallons (41,600 kl) of crude oil bled from the tanker into the waters of Prince William Sound. By the time the leaking stopped, it became the worst oil tanker catastrophe in American history.

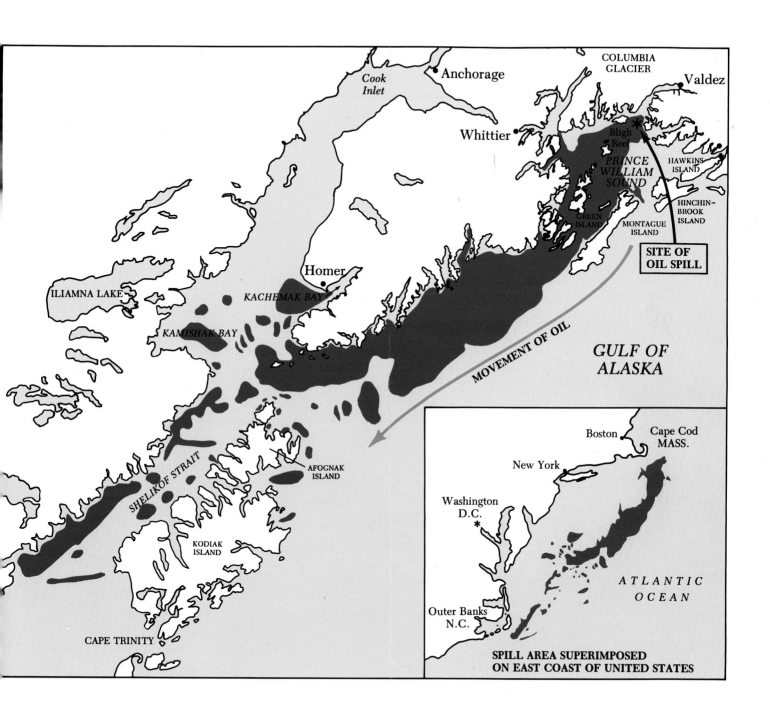

Wind and current drove the oil slick beyond Kodiak Island, 300 miles (483 km) from the accident. (Inset) Had the spill occurred on the East Coast, it would have spanned an area from Massachusetts to North Carolina.

A rescue worker holds an oil-cloaked victim.

2
HARVEST OF DEATH

Prince William Sound, a body of water that connects part of Alaska's southern coast with the Gulf of Alaska, is one of the state's most beautiful regions. At 15,000 square miles (38,850 square km), the Sound is twice the size of the state of New Jersey. Mountains surround the Sound, holding it in a sort of deep cup. Thick forests of spruce grace the mountain slopes. These trees remain green all year, giving the Sound a sharp contrast in winter between the green trees and the white snow.

Dozens of islands—some big, some small—are scattered across the water. In the sunlight, the water resembles the brilliant blue of a jewel. The shorelines are not the white sand of Florida or California beaches. They are mostly rugged rocks, gravel, or gray sand. Some shorelines have no beach at all, just steep rock faces that plunge into the water.

Besides Valdez, the towns of Cordova and Whittier are on the Sound. Many tiny villages also dot the shorelines, and others have been built on the islands. These villages depend heavily on fishing for their survival.

Only one road goes to the Sound, the two-lane highway that ends in Valdez. The Alaska Railroad goes to Whittier, but all other places in the Sound can be reached only by boat or airplane.

The Sound is one of the most environmentally sensitive regions in the world. A huge range of animals and plantlife thrive there.

Prince William Sound supports one of the nation's richest—and most fragile—ecosystems. Above, glaucous gulls dot the craggy hills of Perry Island.

Humpback and killer whales roam the water, looking for a meal of shrimp or fish. And, indeed, the Sound is home to many fish, including halibut, pollock, herring, and salmon. Each spring and summer, about 15 million salmon return to the Sound and the rivers that flow into it. They are returning, after two to six years at sea, to the places where they were bred to lay their eggs. These fish are the main source of income for the surrounding towns and villages. Commercial fishermen await their arrival with excitement. Sport fishermen, too, count the days until the salmon will arrive.

The commercial fishing industry brings in yearly revenues of over $100 million to the Sound.

During the fishing season, canneries open to prepare the commercial fishermen's catch for sale to other states and countries.

Sea lions populate several of the Sound's islands. Sea lions can weigh up to one ton. They come to the Sound in the spring and remain through much of the summer, looking for mates or rearing their young. Seals, too, arrive in the spring. Much smaller than sea lions, they usually weigh less than 250 pounds (113 kg). They come to the Sound to give birth to their pups.

The Sound's most appealing sea creature is the playful sea otter. Otters live in the region year-round. Once they were hunted for their fur, but now they are protected by law. They are an attraction for tourists as well as for the people who live on the Sound. They are superb swimmers and appear to spend as much time playing as they do looking for food. Watching them float on their backs while cracking open a clam or crab or grooming their babies is one of the most popular sights on the Sound.

On land, animals big and small live off the Sound's natural treasures.

The Sound is home to otters, whales, bears, and 5,000 of Alaska's bald eagles.

Bears wander the forests and mountains, searching for food. Deer munch leaves and berries. Marmots, short-legged animals somewhat resembling large squirrels, scurry about feeding on grass, roots, and berries.

Majestic bald eagles dominate the skies. With a wing span of up to 7 feet (2.1 m), they soar above the water and forests with spectacular grace. They feed on fish and small land animals.

Gulls, sea ducks, and Canada geese spend part or all of the year in the Sound. Murres (a seabird that returns to the Sound in the spring) and loons (which breed in large colonies in the Sound) populate the area in huge numbers during spring and summer.

In the springtime, the Sound is waking from winter. Ice and snow are melting. Bears emerge from their hibernation dens. Fish and birds that winter elsewhere begin to return. Prince William Sound is coming to life.

The wreck of the *Exxon Valdez*, however, changed all that. The oil spill turned a time of awakening and beauty into a time of nightmare and death. The Sound awoke on March 24, 1989, to find itself the victim of a disaster unlike anything that had occurred before in the United States.

Alaska's brilliant sunlight reflects off the oil spill's swirling patterns.

*Oil-soaked kelp and sludge streak
the pristine blue waters of the Sound.*

One of the worst parts of the first few hours of the spill is that no one was prepared for it. The oil-spill response plan calls for spill-fighting equipment to be on hand five hours after a spill occurs. In fact, ten hours passed before the Alyeska Pipeline Service Co. got people and equipment onto the water and to the oil. Alyeska is the company formed by the oil companies that own the trans-Alaska pipeline to operate the line. The company also runs the pipeline terminal and is responsible for responding first to any spill.

During those critical first hours of the spill, Alyeska crews worked frantically to get oil-containment equipment out on the water. But little of the equipment was ready to go. A barge that should have had much of the equipment on it sat nearly empty. Oil-containment booms (floating, flexible tubes of plastic used to corral oil) and other supplies had to be found in Alyeska warehouses, dug out, and loaded on boats. Snow buried other equipment. It took hours to get all this material ready.

By daylight Friday morning, the oil slick from the disabled ship had

The eye of the storm: 42 million gallons (159 million l) of the grounded tanker's unspilled oil were pumped into smaller vessels. Twelve days after the accident, the Exxon Valdez was towed to safe anchorage.

spread out for miles. Oil had swept over Prince William Sound like an unstoppable ocean wave, only this wave was thick, black, and deadly.

Tides and wind pushed the oil southwest, toward the heart of the Sound. This movement spared Valdez itself and the Sound's east coast. Still, it wouldn't take long before the oil reached the islands and the west coast, coating their shores with a thick, gooey slime.

Beautiful weather settled over the Sound during the spill's first few days. Steady, brilliant sunlight warmed the air. The water was mostly calm. The fine weather, however, worked against early attempts to control the advancing oil.

Chemicals, called "dispersants," that might have been used to break up the oil require rough seas. Like dishwater soap, these chemicals need turbulence to work best. They need wind and waves for them to spread, foam, and do their job. The chemicals were tested, but the tests failed. The water was too calm.

By the third day, the oil slick covered 100 square miles (26,000 hectares). Its size grew with remarkable speed. Fishermen took to the water in boats. They began using buckets to scoop up the oil off the water and take it to shore. These efforts had only a tiny effect.

Then, on the fourth day, the weather worsened. While calm waters had prevented dispersant use earlier, this time the weather was too stormy. Planes could not fly to spread the chemicals. Winds of up to 50 miles (80 km) per hour pushed the oil around the Sound. The poor weather lasted two days. By the time planes could get in the air again, it was too late. The oil covered an area far too large for dispersants to work.

Attempts to corral the oil with containment booms, which are designed to prevent the oil from spreading, weren't working either. There was too much oil and too little boom.

Workers tried to protect water quality and fisheries by floating containment booms to corral the oil during cleanup operations.

28

APR 13 '89

Containment booms formed a necklace around the Exxon Valdez to contain its leaking oil.

The Exxon Corporation had taken control of the cleanup after the first hours. Fishermen grew angry at the giant company. They said Exxon wasn't doing enough to stop the oil or to clean it up.

Some fishermen took matters into their own hands. They organized attack forces. They gathered as much equipment as they could and set up barriers of containment booms to protect fishing grounds and hatcheries, where millions of fish are raised in pens. Containment booms float on the water, much like a string of logs would. In some cases, the booms worked. In other cases, there was too much oil. It simply washed over the booms.

Exxon hired dozens of boats. Just about anything that could float went out to try and stop the oil. Some of the boats took workers out to set up camps

Workers were hoisted aboard a cleanup boat off Eleanor Island.

Modern technology—helicopters, chemical dispersants, and large quantities of supplies—were not enough to stop the advancing oil slick.

The cleanup effort was a dirty task, staining humans and boats alike.
(Above, left) Thousands of bags of oily debris were later burned in incinerators.
Some of the oil recovered from the waters was recycled;
other debris was shipped to a toxic-waste facility in Oregon.

on islands and shores. Others carried workers out at dawn and returned to Valdez or Cordova at night. It was hard, dirty work. Oil fumes made some workers sick. Oil quickly dirtied any boat or person that got near it.

By the seventh day of the spill, oil had moved out of the mouth of the Sound into the Pacific Ocean. Currents pushed it west, along the state's

The Alaska state ferry **Aurora** *served as a floating hotel for workers fighting the spill. (Facing page) The oil washed up on hundreds of miles of shoreline.*

southern coast. Within a few days, oil threatened the coast of Seward and Kodiak Island, two other major fishing areas.

As days passed, workers built more and more camps. Some were built on shore, but others were built on barges, which floated on the water. Ships and large boats were also used to house workers. These floating living quarters became known as "floatels." In some places, so many people, boats, and equipment moved in that instant towns were created.

All this time, the oil continued to spread. It reached into every corner of some islands. It washed up on hundreds of miles of shoreline. It turned beaches black with slippery slime. Each high tide laid on a new layer of the muck.

The tides also washed up a more terrible toll: dead birds and sea otters. Within a week after the spill, thousands of dead murres, loons, and other birds littered the beaches. One island had 500 dead birds on a 4-mile (6.4 km) stretch of shore. Oil had turned the dead birds into a stiff, black mass.

Some of the birds had drowned. Once the heavy oil got onto their feathers, they could not float. Other birds died from the cold, because feathers contaminated with oil lose their shape and thus their ability to insulate. This left the birds vulnerable to the deadly cold of the Sound's waters. Yet more birds were poisoned when they ate plants that had oil on them. The toxic chemicals in the oil destroys the birds' internal organs or poisons their blood, causing them to weaken and die.

Bald eagles found the birds on the beaches an easy meal. Some of the eagles, though, would themselves die from eating prey contaminated with oil.

The saddest victims to see on the beaches were the murres and loons covered with oil but not yet dead. Many of them were completely black. They struggled to fly, but could not because they were too heavy with oil. Rescue workers captured many of them and took them to bird-care centers for cleaning.

As much as the birds suffered, the Sound's otters may have suffered more. About 10,000 of the playful creatures live in the Sound. Scientists counted almost 500 dead otters in the spill's first days. Many more otters probably died and sank in the water.

Oil ruined the otters' ability to stay warm. Matted with oil, their fur could no longer insulate their bodies from the cold water. Others died from damage to internal organs caused by the oil's poisons. Rescue workers were

Heartbreaking images of death and near-death awaited workers first on the scene. By summer's end, the U.S. Fish and Wildlife Service had counted over 27,000 dead birds.

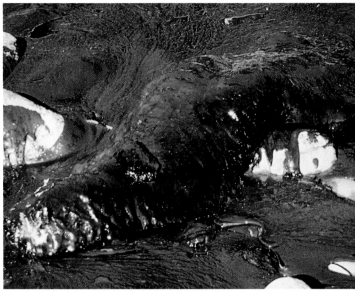

able to capture about 100 oiled otters that were still alive in the first few days. They took them to shore for care and cleaning.

While birds and sea otters were obvious victims of the spill, the effect of oil on the fish and other wildlife was less certain. Whales, for example, seemed to avoid the oil. And even if they got into the slick, they had a thick layer of blubber that would protect them against the cold. Seals and sea lions had the same protection. Still, oil on their food might make them sick or kill them. It would take years of study before scientists would learn the full effects of the oil on these animals.

The effects of oil on salmon, salmon fry (baby salmon), herring, shrimp, and bottomfish would also require study. One thing was certain: Fishermen could not fish in areas contaminated by oil. Therefore, state officials immediately closed all herring fishing in the Sound. They closed many areas to shrimp, crab, and to bottomfish fishing.

Officials closed areas in the Sound and along the state's southern coast to salmon fishing. They even closed some areas reaching up into Cook Inlet, far

(Facing page) Animals that fell prey to the thick crude included fishes, sea otters, and harbor seals. (Below) Sea lions aboard a navigational buoy seek refuge from the oily waters.

from the original spill. These closings hit fishermen hard. Many depend on the fishing season to make the money they live on for the rest of the year.

Fishermen feared the oil could affect salmon fishing for years because salmon fry are particularly sensitive to oil. Normally, the hatchery-raised fry are released in April and spend up to three months feeding near the shore before swimming out to sea. During these months, even small amounts of oil can kill them and reduce the numbers returning in later years.

The oil threatened basic links in the Sound's food chain, such as kelp (left) and mussels.

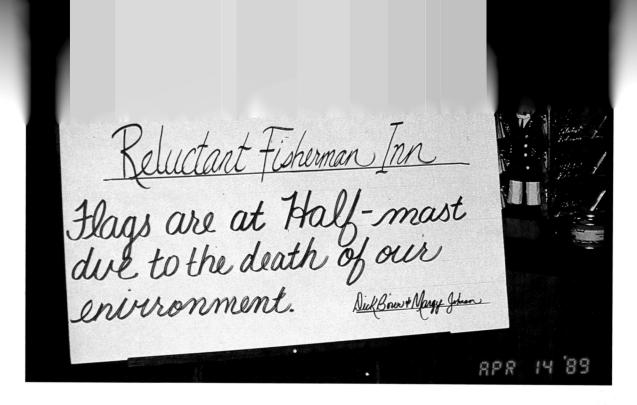

APR 14 '89

The spill's threat to the Sound's salmon, herring, halibut, and shellfish angered and saddened local fishermen.

The Sound's food chain was threatened down to the smallest creatures and plants. Zooplankton and phytoplankton are the tiny animals and plants upon which almost everything else in the Sound depends. Salmon fry and other sea life feed on them, bigger creatures feed on the fry, and so on up to the whales, the biggest animals in the Sound. The oil spill struck at just the time when the plankton were beginning to bloom. Widespread destruction of them could affect sources of food all the way up the chain.

The Exxon Corporation was blamed for the entire disaster. After all, it owned the ship that spilled the oil and employed the captain who was in command. At fishermen's meetings, in the newspapers, and on television and radio, people criticized the company. Exxon promised it would pay to clean up the oil. It said it would pay fishermen who were hurt by the spill. Still, the criticism got worse. People saw one of Alaska's most beautiful areas dirtied with oil. Many people saw their way of life at risk. It made them angry. The anger and the heartbreak of seeing more wildlife succumb to the oil would grow in the weeks ahead.

An oil-soaked otter is recovered on the beach of Knight Island.

3

RESCUE AND RECOVERY

Animal rescue workers from all over the world poured into Valdez, Cordova, and Seward. Some were paid by the Exxon Corporation to work on the spill. Many came as volunteers.

These people quickly set up bird and otter rescue centers. They spent hours out on the water, trying to capture the animals and bring them in for cleaning.

It was a difficult and sad fight. The workers could help only a tiny number of the creatures harmed by the oil. Still, they worked night and day, sometimes without stopping. The animals resisted capture. Despite being almost helpless from oil, they struggled to get away. For every one caught, many more escaped and probably died.

It took hours to clean a single bird. First, workers gently scrubbed it with soap and toothbrushes. They used Water Piks, which squirt tiny jets of water and are normally used to clean teeth, to wash around its eyes. Once washed, the bird would be dried and then, if necessary, washed again. The washing-drying-washing cycle might be repeated three or four times before a bird was clean enough to be released to the wild.

The volunteers and other workers worked fast. They tried to clean and release the birds as quickly as possible. They feared that the longer the birds

43

The Sound is abundant with water-loving birds, such as ducks, murres, cormorants, and grebes. Diving birds, like the loon (top), cease to be waterproof in oily water. (Bottom) An animal rescue worker on Green Island delivers two captive birds to an animal rescue center.

When oiled feathers lose their shape, they lose their ability to insulate the animal from the cold. At bird rehabilitation centers, oil-soaked birds are washed clean, and fed. (Lower right) The birds were quickly released to the wild after treatment because they were vulnerable to disease in captivity.

were kept in captivity, the greater their risk of contracting diseases from contact with humans. But there was also the danger that, once released, these birds would be contaminated again.

Otters presented even more problems than the birds. Bigger and stronger, they were harder to catch. Several even bit people trying to rescue them. They were also more difficult to clean. Workers would spend hours trying to scrub the oil off a single animal. Otters got the same wash-dry-wash treatment as birds.

Plywood cages were built to hold the otters while they were being cleaned and treated. Many of the animals were very sick. They lay helplessly in their cages. Some shivered constantly. Some didn't move at all. The oil's poisons damaged the otters' lungs, stomachs, and livers. Half the otters treated during the first weeks of the spill died anyway. Even the ones that survived could not be released immediately. Veterinarians feared that they would return to oiled areas of the Sound or would carry diseases back into the

Sea otters were also ravaged. Marine mammal rescuers transport a dead otter (left) and another oil-contaminated animal back to shore for treatment.

Sea otters needed much more time than birds to recover. This makeshift hospital (left) for otters in Seward kept its furry patients in plywood cages. (Right) An orphaned sea otter clings to an animal rescue volunteer. Sea otters do not have thick blubber. The oil robs their fur of its insulating ability, making the animals vulnerable to lung, liver, and kidney damage and hypothermia.

wilds. During the first summer, 348 otters were treated at the centers and 226 of them were saved, according to the U.S. Fish and Wildlife Service. The amount of money spent on the otter rescue program came to about $80,000 per animal.

Meanwhile, oil continued to roam the Sound. Much of it began to change into a thick, puddinglike mixture called "mousse." Some of it mixed with seaweed and other plants and formed thick tar balls that floated like big chunks of peanut butter.

Whatever form the oil took, though, all of it proved extremely difficult to pick up off the water. The sheen spread over such a vast area that boats and equipment were able to do little good. "Skimmer" boats, with conveyer belts that dipped into the water, attempted to scoop the oil off the water. Other boats tried using suction hoses. But chasing the oil around the Sound was like trying to clean soap suds out of a kitchen sink with a spoon. Even a giant skimmer hired from the Soviet Union, the *Vaydaghubsky*, had trouble handling the thick oil. The Russians finally gave up and went home.

On shore, attempts to pick up the oil were just as painstaking and difficult. The slippery, black slime was like a layer of hot fudge over the rocks, gravel, and sand. Some of the oil had sunk deep into the ground. And some of it would wash off with an outgoing tide, only to return when the tide came in again.

At first, the Exxon Corporation hired people to walk the beaches cleaning the rocks by hand. The workers picked up each stone on a beach, wiped it as clean as they could, then put the rock back where it was. It was a

(Below) Workers clean a large skimmer unit, which works like a vacuum cleaner, to suck up floating oil. (Facing page) The black tide left its mark on Treasure Cove. Cleanup units painstakingly tried to clean several coastlines, one rock at a time.

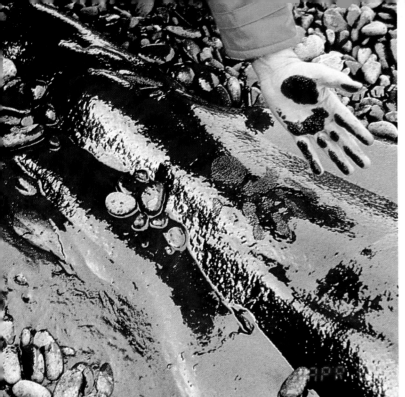

(Left)The black slime spread like a layer of hot fudge over the rocks. (Right) This picture contrasts three hand-cleaned rocks against an oiled beach.

painfully slow process, and it didn't do much good. Tides would simply cover the rocks with oil again. And the hand-cleaning method did nothing about the oil that had sunk in the ground.

As more shore-cleaning equipment became available, the workers started trying to clean the beaches with water. This method involved spraying the shore with high-pressure streams of water, much like cleaning a driveway with a water hose. The hoses washed oil and debris off the beach into the water, where it was corralled with floating booms and scooped up by skimmers.

But the washing method, too, had problems. It was slow. It didn't get oil below the surface, and it didn't get off all the oil on the surface. Even after repeated spray-cleaning, beaches were still stained black with oil.

More important, the washing method knocked loose and killed many small creatures like mussels and snails. This process became the most widely

50

Workers also sprayed oil-contaminated beaches with water.

used means of cleaning rocks and shorelines, but scientists agreed that at some point the force of the water would cause more harm than if the beaches were just left alone.

The cleaning method that showed the most promise was the speeding up of a natural process. This was done by applying a fertilizer to the oiled shorelines that increased the growth of "oil-loving" bacteria. These bacteria occur naturally. Over the years, they would eventually eat the oil anyway, although it would take twenty or more years. The fertilizer increases the bacteria's growth and the speed with which they would eat the oil. However, the method is new and was used cautiously in Prince William Sound. Even after tests showed good results, approval for wide use of the fertilizers was slow in coming.

All of these cleanup efforts required thousands of men and women. As summer wore on, the Alaska oil spill's work force numbered nearly 11,000.

Valdez was flooded with more people than the little town could handle. Hotels quickly filled. Some people had to live in tents; others built shelters out of any scrap materials they could find.

Restaurants, hotels, gas stations, grocery stores, clothing stores, and other local shops did a booming business. Cleanup workers were paid well. Most received about $16.50 per hour. They had a lot of money to spend. All those people and all that money turned Valdez into a busy, crowded place, nothing like the quiet town it was before. For all the damage and sorrow the spill caused, some people still made a lot of money from it.

As for the crippled *Exxon Valdez,* the vessel left Prince William Sound on June 23, exactly three months after she had departed Valdez on her original voyage. During those months, the oil that didn't spill was transferred from the *Exxon Valdez* to another tanker. Tugboats towed the damaged vessel to one of the Sound's islands, where underwater divers and other workers made what repairs they could to the vessel's torn hull. Then the tanker was linked to two other tugboats, which towed her away from Alaska.

Worldwide attention brought ships such as the Soviet skimmer, the **Vaydaghubsky** *(docked in the background), to take part in the cleanup effort.*

After a year of repairs, the **Exxon Valdez** *was back in the water with a new name, the* **Exxon Mediterranean.** *(Right) The vessel's hull towers over a man, giving us a perspective of the great size of the tanker.*

Her destination was San Diego, California, where major repairs would make her seaworthy again.

The tanker left behind more than 1,000 miles (1,610 km) of oil-slimed shoreline in Prince William Sound. If the oiled areas of the Sound were laid against the eastern coast of the United States, they would reach from Massachusetts to North Carolina. The tanker left behind a trail littered with thousands of dead birds and sea animals. It left a ruined fishing season for fishermen. And it left many questions about whether the Sound would recover from the spill.

By the end of the first summer after the spill, rescue workers had counted about 30,000 dead birds. Wildlife biologists estimated, though, that this number is only 10 to 30 percent of the toll, meaning that between 90,000 and 270,000 birds have probably died and disappeared in the waters of the Sound. They also counted 1,016 dead sea otters. The toll in both otters and birds could increase as the years pass. Nobody can say for sure how high the numbers will go or how long the oil will continue to harm wildlife.

53

The spill marred the beauty of the Alaskan landscape.

4
AFTERMATH

Many questions about the recovery of Prince William Sound cannot yet be answered. For the most part, all people can do is clean what can be cleaned, then let nature takes its course. Much of the oil quickly evaporated immediately after the spill. Some dissolved in the water. Workers in boats and on the beaches cleaned up more. But much of it remained in the water or on the shorelines. It is up to the natural oil-eating bacteria to do away with what remains. Scientists say this will occur, but it will take time.

As the oil disappears naturally, the effects of the spill on fish and other sea life will decrease. How long this will take is unknown. Other spills in different parts of the world have shown that water and land will eventually become clean of oil. In 1970, a major tanker spill off the coast of Nova Scotia produced evidence that the oil will gradually disappear. Six years after that spill, ocean waves and weather had greatly reduced oil on the shoreline. One of the biggest tanker spills in the world occurred in 1978 in the English Channel. Human efforts and natural processes returned the region to its original state about eleven years later.

These other spills have shown that calm waters recover the slowest. Here, Prince William Sound has an advantage. Raging storms are common in the Sound, particularly in winter. The natural weather of the area should help wash the oil from the Sound. However, coves and bays protected from the weather will take longer to become clean. Oil could remain in these areas for years.

In addition to the physical destruction the spill caused in Prince William Sound, the accident changed many people's view of oil development in Alaska and elsewhere. People demanded that the oil industry be watched more closely to assure that such a spill will not happen again. Investigations into the spill's causes and effects began immediately in Alaska and Washington, D.C.

From the viewpoint of the oil industry, the spill could not have come at a worse time. The industry was trying to obtain approval to look for oil in other parts of Alaska, particularly in northern Alaska's Arctic National Wildlife Refuge and in Bristol Bay off the state's southwest coast. But the damage caused by the *Exxon Valdez* strengthened the arguments of those opposing new development.

Alaska's government quickly passed new laws and created new oil industry regulations. These changes call for better plans to deal with oil spills, stricter procedures aboard tankers to assure safety, and more effective ways of guiding tankers when they are close to shore. Other laws increase the penalty an oil company will have to pay if another spill occurs.

The state also filed a multibillion-dollar lawsuit against the Exxon Corporation. The suit demands that Exxon pay for all the damage caused by the spill. The suit is still in progress as this book is being completed.

Furthermore, federal authorities filed criminal charges against the company. The most serious alleged offense, a felony, claims that the company violated federal laws requiring competent crews aboard oil tankers. In other words, the government claims that Exxon failed to make sure a qualified crew operated the *Exxon Valdez*.

The other charges claim the company committed the misdemeanors of dumping oil, polluting the water, and unlawfully harming wildlife. Exxon denied all the charges.

(Left) Alaska governor Steve Cowper has asked that the oil industry develop stronger measures to prevent and combat future oil spills. (Right) Joseph Hazelwood, former captain of the Exxon Valdez.

Both sides, the federal government and Exxon, attempted to settle the case out of court. If the company is convicted on all the charges, it could face fines of more than $700 million. Like the state's civil lawsuit, the criminal case is still in progress as this book is being written.

Another case, this one against *Exxon Valdez* Captain Joseph Hazelwood, did end exactly one year after the original spill. On March 24, 1990, a criminal court judge in Anchorage, Alaska, sentenced Captain Hazelwood to spend 1,000 hours cleaning beaches in Prince William Sound. The day before, a jury had convicted Captain Hazelwood of the misdemeanor of negligent discharge of oil. The jury found the captain not guilty of three other charges, including one of operating a ship under the influence of alcohol.

Superior Court Judge Karl Johnstone had the authority to send Captain Hazelwood to jail for ninety days. But he chose instead to make him help clean up the spilled oil. And in another part of his unusual sentence, Judge Johnstone ordered the captain to pay partial restitution of $50,000 for the

A scientist collects fish from the Sound in an effort to study the spill's long-term effects.

damage that the *Exxon Valdez* caused. Captain Hazelwood's lawyers said they would appeal his conviction.

Since the accident, scientists from all over the world have gone to Alaska to study the effects of the spill. In a way, Prince William Sound became a huge laboratory for these scientists. They hope to learn more about what oil can do to land, water, and wildlife. They hope to learn which cleanup methods work best. The amount of our knowledge about the ecological effects of oil spills is small. Scientists say the Prince William Sound spill can add much to this knowledge. The information will be useful in dealing with future spills.

It is a sad fact that accidents happen. But as long as oil has to be moved by ships, oil spills will occur. In fact, other major spills have already occurred in the United States since the Prince William Sound disaster. They were not anywhere as big as the Alaska spill, but they still point to the dangers of transporting oil by sea. That danger will exist as long as Americans need gas for their cars, airplanes, and factories.

Chemists report that crude oil is both toxic and long-lasting.

In cold waters, residue from the spill may take years to decompose. The oil could be absorbed into the food chain through plankton and other tiny organisms over time.

Meanwhile, people who live on Prince William Sound and others who care about it can only wait and watch. They wait to see if the playful otters will recover from the oil's deadly effects. They wait to see if the birds will keep coming back to an area that killed so many of their kind. They wait to see if the black stains on the Sound's beaches will ever wash off.

Experts predict that the Sound will recover. Even some of nature's most fragile creations have a wonderful resiliency. Still, accidents like the *Exxon Valdez* spill do damage, and it is our responsibility as caretakers of the earth to guard our environment from such disasters.

*Perhaps over time nature's resilience will heal the
oil-scarred wilderness of Alaska's Prince William Sound.*

SELECTED READING LIST

Books

Hanrahan, John and Peter Gruenstein. *Lost Frontier.* New York: W.W. Norton & Co., 1977.

Mickelson, Pete. *Natural History of Alaska's Prince William Sound.* Cordova, Alaska: Alaska Wild Wings, 1989.

Magazines and Newspapers

"Alaska After Exxon." *Newsweek,* Sept. 18, 1989, pp. 50–64.

"A Special Report from Alaska: Oil, Water, and Wilderness." *Wilderness,* Summer 1989, pp. 3–5, 64.

"Biologists Track a Slaughter." *Anchorage Daily News,* July 13, 1989.

"Environmental Politics." *Newsweek,* April 17, 1989, pp. 18–19.

"Joe's Bad Trip." *Time,* July 24, 1989, pp. 42–47.

"Paradise Lost." *Alaska,* June 1989, pp. 21–35.

"Rescuers Create a MASH Unit for Hundreds of Stricken Animals." *The New York Times,* April 4, 1989.

Special Oil Spill Issue. *Alaska Fish and Game,* July–August 1989.

"Tapes Reveal Captain Tried to Free Valdez." *Anchorage Daily News,* April 25, 1989.

"The Events After the Grounding." *Anchorage Daily News,* May 14, 1989.

"The Events Before the Grounding." *Anchorage Daily News,* May 14, 1989.

"The Two Alaskas." *Time,* April 17, 1989, pp. 56–66.

"Time Will Tell How the Sound Recovers." *Anchorage Daily News,* May 6, 1989.

"What Went Wrong and Why." *The New York Times,* April 15, 1989.

"When Did the Ship Hit Reef?" *Anchorage Daily News,* May 21, 1989.

"When Exxon Valdez Ran Aground, the Crew Didn't Know What to Do." *Anchorage Daily News,* May 18, 1989.

INDEX